Austin Craig Apgar, Addison Emery Verrill

A Key to the Mollusca

Verrill's Systematic Catalogue of the Invertebrates of Southern New England and the Adjacent Waters

Austin Craig Apgar, Addison Emery Verrill

A Key to the Mollusca
Verrill's Systematic Catalogue of the Invertebrates of Southern New England and the Adjacent Waters

ISBN/EAN: 9783337330798

Printed in Europe, USA, Canada, Australia, Japan

Cover: Foto ©berggeist007 / pixelio.de

A KEY

to the

MOLLUSCA

given in

VERRILL'S SYSTEMATIC CATALOGUE,

of the

INVERTEBRATES

of

SOUTHERN NEW ENGLAND,

and the

ADJACENT WATERS.

A. C. Apgar

Trenton, N.J.

Hektograph Print.

1880.

PREFACE

As far as the author knows this is the first key of any of the mollusks of the United States published.

He claims originality only in the key portion, and in the arrangement of that key so as to show the classification of the species into orders and families at each step. He has made use of all the works at his command in the wording of his descriptions.

He hopes that it will enable any one to trace any good specimen, that is moderately mature.

The dimensions are all given in mm and refer to mature specimens.

The figures on page 8, the Fig. [illegible] page [illegible], and [illegible] page [illegible] are from Sea Fisheries [illegible] Coast of New England. Figs. 2–6 page [illegible] page [illegible] are from Binney's [illegible] Invertebrates of Mass. Fig. 1 page [illegible] 1 page [illegible] are from Morse's First [illegible] in Zoölogy. Figs. 1–2 page [illegible] are original.

CONTENTS.

The derivation of the names are given in parenthesis.

MOLLUSCA.

Soft-bodied, unsegmented animal often protected by a shell; the diges-tive system includes stomach, intes-tine and anus; the nervous system normally consists of three pairs of ganglia, but in the Scaphopoda, Tunicates [illegible] Polyzoa there is only one ganglion. The heart has two or more chambers, but in the Tunicata it is reduced to a simple tube, and the Polyzoa have no hearts.

The integument of the body, continuous or divided into two lobes, is called the mantle. The mantle secretes the shell which is rarely absent.

In the mouth of the Pteropoda, Gastropoda and the Cephalopoda there is an organ armed with a ribbon of teeth; this is the radula, erroneously called tongue.

I Animal with a shell; single.

A Shell univalve.

1 Shell divided across into several chambers, curved into a spiral the curves separated. (Only one example in the section covered by this vol.)

Class Cephalopoda, fam. Spirulidae
p. 12.

2. Shell undivided, usually spiral, often operculated; animal with a disk [illegible] used for locomotion.

Class Gastropoda.
P. 13

3. Shell [illegible], pen like in form; animal with 8 or 10 arms around head provided with

MOLLUSCA.

[illegible] like suckers; eyes large.

Class Cephalopoda. [illegible]

A. Shell thin straight, [illegible] spiral; animal with wing or [illegible] like expansions on the sides [illegible] head or neck by which it swims [illegible] ely in the ocean.

Class Pteropoda, Order Thecosomata
Page [illegible]

B Shell bivalve.

1. Shell exactly equilateral, but inequivalve, one above and the [illegible] below the animal, no branchia; mouth with two long cirriferous arms [illegible] example in the region)

Class Brachiopoda.

2. Shell never exactly equilateral, usually equivalve, sometimes inequivalve, usually provided with a hinge, not with hinge then often provided with accessory valves.

Class Lamellibranchiata.
Page 2[illegible]

C Shell multivalve,

composed of eight transverse, imbricated plates; animal with [illegible] creeping foot.

Class Gasteropoda, [illegible] Polyplacophora [illegible]
Page 11

II Animal without shell, compound.

[illegible] compound fixed minute animals provided with tentacles around mouth

Class Polyzoa.

MOLLUSCA.

III Animal without a shell; simple.

A Free swimming animals with distinct head and provided with two wing or fin like expansions attached to the sides of the neck.

Class Pteropoda, Order Gymnosomata

Page [illegible]

B Animal with head and provided with a large locomotive disk or foot, some provided with wing like expansions, but these are attached to the sides of the body by their lower edge. (See slugs)

Class Gasteropoda, [illegible]

Page [illegible]

C Headless, transparent, jelly like, free swimming animals, with simple ribbon shaped branchiæ. Solitary generation of

Class Tunicata, Order Biphora.

D Fixed animals provided with a leathery elastic integument, having two prominent apertures, simple [illegible]

Class Tunicata Fam. A[illegible]

IV Animal [illegible] a shell [illegible]

A Headless, transparent, jelly like, free swimming animals with simple ribbon shaped branchiæ, [illegible] together in a double [illegible] generation of

Class Tunicata, Order [illegible]

B Fixed animals protected [illegible] in a [illegible] jelly like [illegible] for each individual, [illegible] in many cases an oral aperture [illegible]

each individual and an atrial [illegible] for a distinct group of them [illegible] without the power to project any part beyond the enveloping sac.

Class Tunicata, Order Ascidioida

Page

C Compound fixed animals, each living in a cell of a plant like organism; cells not in communication; animals provided with a row of tentacles around mouth; the alimentary canal suspended in a double walled sack. The investing membrane is corneous or fleshy and usually highly charged with carbonate of lime.

Class Polyzoa.

CEPHALOPODA.

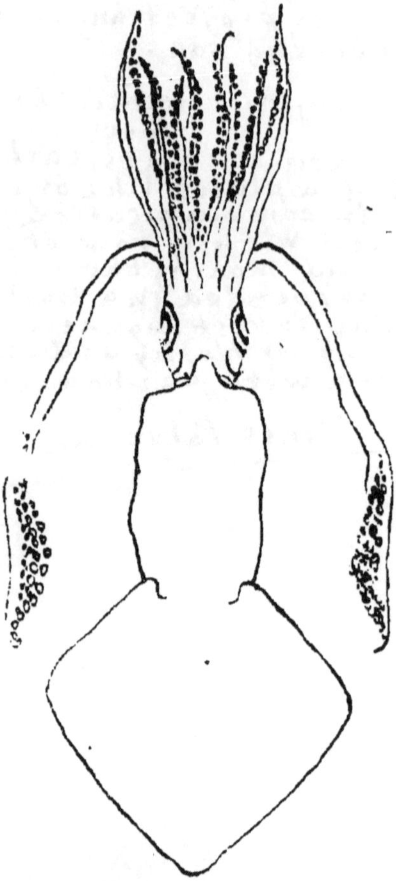

Loligo pallida 1/3 nat. size.

Pen of L. pallida 1/2 nat. size

CEPHALOPODA.

Free swimming oceanic mollusks with a distinct head and long arms placed around the mouth. Eyes large and much like those of vertebrates.

Arms eight or ten, all furnished with [illegible] suckers [illegible] without an external [illegible].

I No external shell, but with an internal straight or curved cartilage which extends the whole length of the body; provided with an ink bag.

Lanceolate pen with three ribs, one central and two marginal; fins rhombic posterior caudal, eyes with lid [illegible] upper [illegible].

Color vivid and beautiful, passing from a brilliant red to a deep [illegible] blue upon the back, the region of the eyes finely tinted with yellow; [illegible] of arms [illegible].

Ommastrephes illecebrosa.

II Internal pen [illegible] with thin edges; fins rhombic posterior dorsally, eyes smoothly covered with the skin membrane.

Caudal fin about [illegible] broad [illegible] [illegible] broadly at all round [illegible] tentacular arms as long as the [illegible] when extended. Length from base of arms [illegible], body [illegible], caudal fin 70, width [illegible] tentacular arms 150. General appearance [illegible] translucent, not very thickly [illegible] with spots.

Loligo pallida.

CEPHALOPODA.

Lateral angles of caudal fin quite rounded; darker color, and thickly covered with spots.

Loligo pealii.

II The internal shell, but with an internal coiled chambered shell which is calcareous and is coiled in one plane.

Shell white and pearly and with spire ... three turns which do not touch each other like a ram's horn; thrown up after storms on the shores of Nantucket.

Spirula fragilis

GASTROPODA

Gasteropoda. Land or water mollusks generally encased in a univalve shell. Locomotion effected by the ventral disk or foot. A distinct head in nearly all, with one or two pairs of tentacles. Diœcious or hermaphrodite. Eyes two or none.

Note, — wh = whorls, ap = aperture. Nos. representing distance are all in m.m. At the end of the descriptions of snails the Nos. are for length, width, angle of spire, No. of whorls, and ratio of aperture to length of shell.

1 Shell spiral; whorls together; dextral, A; operculated; C

A Margin of shell ap. notched or produced into a canal.

Shell with a decidedly produced anterior canal; ap. entire posteriorly; foot broad.

Shell large; canal more than 30 long; ap. longer than spire.

Spire with a broad deep channel at sutures; 180, 80, 70, 6, 3/4. Sycotypus canaliculatum.

A series of tubercles around the wh. of spire. 190, 75, 85, 6, 3/4. Fulgur carica.

Shell medium; ap. about = the spire

Canal more than 15 long, gently curved; ap. polished white within. 70, 37, 50, 8, ½.

Neptunea curta.

Spire without rib like undulations; canal & long; drab wooly or velvety epidermis. 20, 11, 30, 6, 2/5. Neptunea pygmaea.

Spire with 10 to 12 rib like undulations; wh. rounded; 25, 14, 45, 5, ½. Urosalpinx cinerea.

Spire = 10 to 600? equidistant thread like revolving lines; ap. within bright reddish brown.

17, ?, ?, 6, ?. Ptychatractus ligatus.

? ... angular and pointed at the crossing of the 11 elevated ribs; canal slender.

25, 14, 60, 5, 4/5. Eupleura caudata.

B Canal not much produced; ap. notched; 2/5 or m inner lip without teeth or callosity.

More than 25 long.

1.

GASTEROPODA

... obliquely waved elevated ribs; ap. yellowish. 69,37,67, 6,1/2 Buccinum undatum.
Ovate pointed solid; many coarse revolving ridges; very variable in color, from white to chocolate, uniform to banded, and surface from rasping to smooth, [illegible] to raw, Purpura lapillus.

[illegible]

Body wh. upper half marked by 17-20 [illegible] ribs; suture not very distinct; whole surface covered by revolving [illegible]

[illegible]

Anachis [illegible]

Body wh. with [illegible]

Anachis [illegible]

[illegible] and [illegible], outer lip [illegible]; body wh. with 13 oblique [illegible] ribs; flesh colored. 1-,6,30,6,% Bela harpularia,
with about 20 longitudinal oblique folds, numerous revolving raised lines; suture well defined; outer lip thin crenulated. [illegible]

[illegible]

[illegible] and the [illegible] the rib in width; numerous inconspicuous revolving lines; outer lip sharp. 18,5,40,6,%
Bela pleurotomaria

[illegible] with a callosity or blunted [illegible] plait; ap. ovate; [illegible]. Suture distinctly marked; surface with numerous unequal revolving lines crossed by minute lines of growth and oblique folds; very common on sand and mud; dark colored. [illegible]

Ilyanassa obsoleta.

Suture shouldered; whole surface covered with a network formed by many longitudinal and about 10 revolving lines on the body wh. operculum triangulated and [illegible] of its edge serrated. 18, 8, 47, [illegible] Tritia trivittata.

Suture not distinct; outer lip with [illegible] within; [illegible] white, [illegible]

[illegible]

GASTEROPODA

Body wh. with 12 rib like folds and crossed by 10 or more elevated revolving thread rendering the ribs nodulous; spire with ribs and revolving lines distinct; suture not distinct. 6, 3, 40, 6, ⅔. Bela plicata.

Closely covered with almost microscopic revolving lines; outer lip sharp, slightly everted, smooth within; suture faintly impressed but distinct; white tinged with rose. 7, 4, 35, 6, ⅔. Astyris rosacea

One revolving line below suture [illegible] around beak; reddish-brown with crescent shaped yellowish spots on body wh. 5, 25, 43, 6, ⅓. Astyris lunata.

No revolving lines on beak, slightly striated longitudinally. 5, 25, 43, 5, ⅓. Astyris zonalis.

Two revolving ridges with distinct groove between around the wh. 8, 4, 40, 6, ⅔. Pleurotoma bicarinata.

Suture distinct but shallow, undulated; body wh. with about 11 prominent longitudinal ribs separated by wide concave spaces; wh. angulated at the middle and decidedly flattened below suture. 6.5, 3, 41, [illegible]. Mangelia cerina

Shell turreted, 15 or less long; ap. ¼ or less of the length of the shell.

— Canal deep, short, slightly curved.

Reddish-black; 3 series of granules on the lower wh. caused by 3 revolving lines and about 20 ridges; ap. nearly circular. [illegible], 10, ⅓. Cerithiopsis [illegible]

Irregularly granulated surface; suture abruptly and sharply defined, wh. flattened. Beak short, twisted, wrinkled. 12, 3, 20, 16, [illegible]. Cerithiopsis emersonii.

Regularly ridged surface like the thread of a screw, [illegible] and [illegible] transversely [illegible] and striated. 12, 3, 18, 11, [illegible] Cerithiopsis terebralis.

Canal a mere oblique fissure or notch.

Six revolving lines and 20 ribs on the two lower wh. making a granular network over the surface 7, 2, 25, 1, 1/4 Bittium nigrum.

15 Margin of shell ap. entire ?

Shell globular of few wh. with a small spir[illegible] [illegible] lipuscate, umbilicus distinct, either [illegible] [illegible] by callus, ap. [illegible] or more.

[illegible]ate; lines of growth distinct; [illegible] well marked; umbilicus large and simpl[illegible] [illegible], 50, 120, [illegible]. Lunatia heros.

Many young specimens are found [illegible] [illegible] in the [illegible]

[illegible]

[illegible] triseriata.

[illegible] of growth [illegible] [illegible] umbilicus [illegible]

8, [illegible] Lunatia immaculata.

Umbilicus deeply grooved, and partially [illegible] [illegible] covered with a chestnut coloured callus which extends [illegible] along the [illegible] lip. 49, 26, 104, 5, [illegible] [illegible]

[illegible] callus along the inner lip [illegible] into the umbilicus nearly filling [illegible]

5, 0, 90 [illegible]. Natica pusilla.

Small cup shaped or [illegible] [illegible] species [illegible] [illegible], ap. rounded-ovate, [illegible] [illegible] [illegible] or [illegible], slightly [illegible] the somewhat flattened.

[illegible]; plain revolving [illegible] [illegible] [illegible]; ap. obliquely [illegible] [illegible] broadly flattened [illegible] [illegible]

[illegible] yellow [illegible]

5, 10, 70, [illegible]. Littorina [illegible]

Globular-ovate, [illegible], shining, spire [illegible] [illegible], white, yellow, orange, [illegible] brown, [illegible] [illegible] [illegible], animal [illegible] orange [illegible] and exceedingly [illegible]

[illegible] Littorina [illegible]

[illegible]-ovate; surface [illegible] revolving striated [illegible] [illegible] and thick [illegible]

[illegible]

O.S.O.F.

GASTEROPODA.

PROSOBRANCHIATA. Rhipidoglossa. Trochidae.

3 Conical; ap. circular, pearly within. Umbilicus large and deep; wh. convex and rendered angular by a prominent revolving ridge; lip simple sharp; operculum horny nucleus central, 12, 9, 95, 5, 2/5.
Margarita obscura.

4 Globose or conical, thin; ap. semi-lunar; inner lip oblique; outer lip sharp flattened; umbilicus a lengthened groove along the pillar

Taenioglossa. Littorinidae.

Ovate-conic; wh. encircled by 4 or 5 purplish brown bands and numerous minute undulating lines; suture fine deep; ap. nearly orbicular; inner lip white, flattened and excavated by a smooth, crescent shaped groove terminating in an umbilicus 13, 8, 50, 5, ½
Lacuna vineta

Globular-ovate; ap. semi-lunar, oblique; umbilicus large and deep. 5, 6, 95, 3½, ¾.
Lacuna neritoidea.

Rissoidae.

5 Ap. ⅓ the length of the shell, ovate-conic; spire obtuse at top; suture distinct; wh. 5–6; umbilical chink.

Littorinidae.

Usually coated with a dark green pigment or minute vegetable; animal blackish; thin; common on sea weed about high water mark
4, 25, 25, 5, ⅓ Litteronella minu[illegible].

Rissoidae.

Wh. convex, covered with regular microscopic lines revolving around the shell; suture deep; light yellow horn color; ap. oblique, ovate, angular behind. 4, 17, 25, 6, ⅓. Rissoa aculeus.

Wh. convex; 3 revolving lines on upper whorls; ap. ovate; fuscous. 27, 12, 20, 5, ⅓. Rissoa exarata.

6 Turreted shells, 3 or more times the length of the ap.

Turritellidae.

—Wh. crossed by very distinct elevated ribs; ap. ovate, the margin entirely united; lip continuous, reflected; umbilicus none.

16–18 delicate ribs, not crossing the suture, no revolving lines between; white.
12, 5, 30, 8, ⅕. Scalaria lineata.

17–20 ribs, with the sp[illegible] between marked with numer[illegible] revolving lines; white 12, 5, 30, 8, ⅕. Scalaria multistriata.

GASTEROPODA.

... rounded, flattened, oblique white ribs; intervening spaces marked by 6-7 coarse, ... equidistant revolving ridges; ap. nearly round boardered by by a rib; bluish-white. 24, 10, 12, 10.1/4

Scalaria granlandica.

— Wh. without decidedly elevated longitudinal ridges,

Columella plaited; ap. ovate; lips disunited posteriorly; pillar with a tooth like fold; operculum horny; tip of spire usually obtuse; less than 1 long. Genus Odostomia.

Shell plain and horny; suture distinct.

Tip decidedly blunt; wh. flattish; umbilicus none; 6, 14, 15, 8, 1/4. O. producta

Tip bluntish; sub-umbilicated, ... sometimes seemingly double by a ridge; tooth like fold of pillar sometimes well within the shell. 6, 14, 25, 6, 1/4. O. fusca.

Ivory white, rather solid.

Suture well defined; wh. convex; pellucid 5, 16, 28, 6, 1/4 O. dealbata

Suture slightly but sharply depressed, in which there are about 5 revolving lines of which one above and two below are most distinct; ap. acutely angular above. 6, 2, 23, 8, 1/7. O. trifida.

Glossy, translucent; wh. convex with numerous ridges or folds crossed by three equidistant revolving lines giving the surface a ... appearance except lower half of body wh. which has revolving lines only. 4, 18, 30, 6, 1/3. O. seminuda.

Dusky or greenish.

Dusky; 4 ... impressed revolving lines; acute at apex. 5, 2, 22, 6, 1/7. O. impressa

Surface light green under a brownish epidermis; lower wh. 1/2 length of shell; pillar lip bluish white; outer lip ... simple; umbilical chink; revolving lines below suture whitish. 5, 25, 15, 5, ...

O. [illegible]

GASTEROPODA.

†† Columella straight, simple, without plait; aperture oblong, sub-quadrate or ovate

A reversed wh. at the apex. G. Turbonilla

Wh. slightly convex; 25 ribs; 14 revolving lines looking like 7 because they are in pairs; suture well defined. 6, 2.5, 12, 9, 1/4. T. interrupta.

Wh. well rounded; suture rather deep; numerous ribs not so broad as the interspaces; about 5 revolving lines on the upper whorls, interrupted on the ribs. 5, 1.5, 20, 10, 1/4. T. elegans.

Wh. moderately convex, flattened in the middle; 25 ribs on lower wh. 6 revolving lines on the upper whorls which do not cross the ribs; apical wh. minute; shell obelisk shaped. 4, 1.5, 20, 8, 1/4 T. areolata.

Large smooth reversed apical wh.; wh. slightly convex, flattened; 20 ribs on lower wh.; revolving lines very minute numerous; body wh. with two bands of pale brown. 4, 1.5, 22, 6, 1/4. T. costulata.

Reversed wh. minute; white; very acute; 16-18 ribs on lower wh.; 2 upper wh nearly smooth. 4.5, 1, 15, 10, 1/4. T. stricta.

20 ribs on body wh., interstices deep and appearantly smooth; up. round-ovate. 4.5, 1.25, 18, 10, 1/4. T. equalis.

††† Apex acute; animal and shell [illegible]; suture inconspicuous; wh. flattened marked with light brown transverse band; variable shell white shining; animal hyaline. 6, 1.5, 20, 12, 1/4. Eulima oleacea.

7 Shell discoidal; ap. dilated; concavely umbilicated beneath; all the wh. seen from the under side. 1, 1.3, 1.6, 3, 2/3. Skenea planorbis.

Cerithiadæ

II Shell spiral; wh. together. Sinistral, ; operculated.

Turretted, granulated. 5, 1.2, 23, 1.2, 1/7. Triforis nigrocinctus.

Thin, transparent, ovate-globose, 2.5, 2, [illegible], 7, [illegible]

See No. 2. Page 27.

C..O.F.

GASTEROPODA.

III Shell spiral; wh. separated operculated.

Shell conic tubular; unequal striæ run the length of the tube; Spiral portion 15 long, the rest continued indefinitely, sometimes 200 long; diameter of ap. 6.
Vermetus radicule

IV Shell tubular, merely curved, very minute; operculated.

About 25 strong encircling ribs. l. 2·5; dia. ·6,
Cœcum pulchellum.

Ridged lengthwise of the shell. l. 2·5 dia. ·6.
Cæcum cooperi.

V Shell not appearantly spiral but flat or concave.

1 With an internal, usually horizontal partition or diaphragm.

Oblique-convex oval, partition a[illegible]ssed to one side; apex prominent turned to one side; convexity moderate but different according to the object on which it adheres; common. l. 35, w. 25. diaphragm ½ of ap. Crepidula fornicata.

Ovate flat; apex acute, terminal; diaphragm convex; found inside shells. l. 30, w. 22.
Crepidula plana.

Shell very convex; apex terminal separated from body of shell; diaphragm convex, less than ½ of ap.; inside brown except edge of diaphragm. l. 11, w. 6·5. Crepidula convexa.

Shell sub-conic, oval; apex central; numerous radiating lines; diaphragm triangular fastened by one side, the free point nearly correspond, with the inner apex of the shell. l. 22, h. 18. Crucibulum striat[illegible]

2 Without internal partition.

Shell basin shaped, oblong, oval, thin; apex obtuse; surface finely checkered with

alveus

growth other
2.12 w.7.
Trachydermon ruber

GASTEROPODA.

[illegible]

[illegible] No. 127 68 6, 3/4 Melampus bidentatus

[illegible]

[illegible] 12, 60, [illegible] Melampus [illegible]

VIII Animal without [illegible], branchiae [illegible], [illegible] concealed in the [illegible] or [illegible] length [illegible]

1 Animal [illegible] with [illegible] [illegible] with [illegible] angles, [illegible] tentacles; [illegible] branchiae tufted near the posterior end on right side in groove between mantle and foot; dark brown above, white beneath. 75, 5. Doridella obscura.

2 Branchiae unbranched, club form, a single line on each side of the body.

Slender, gradually tapering backwards, convex above; foot [illegible]; tentacles [illegible]; branchiae [illegible] side; pale rose color dotted with dark brown. 1?, 1?. Doto coronata.

Lance-linear, colorless with a zigzag olive-colored stripe along back connecting the obovate branchiae, 4-? on each side; dorsal tentacles long, simple, blunt,

Eolidae. Hermaeidae.

GASTEROPODA.

oral ones short pellucid; tail pointed
6, 1. *Tergipes despectus.*

Linear; branchiæ 7 large and 8-10 small ones on each side, the large ones are much enlarged near the end, and with the internal dark parts look like crosses.
Hermæa cruciata.

3 Branchiæ simple unbranched, situated in double oblique cross lines on each side of body, club shaped.

Lanceolate, widest 1/3 from head; drab color, back with a carmine line margined with silvery dots between tentacles and each tuft of [illegible] branchiæ; tail pointed, silvery; branchiæ in 5-7 double rows on each side. 36, 6.
Montagua pilata.

Lanceolate; dorsal tentacles longer than oral ones; light yellow tinged with pale orange; branchiæ 5 double rows on each side about. [illegible] 3.
Montagua gouldii.

Lanceolate tapering to a point; head rounded; oral tentacles longer than dorsal ones; branchiæ about 6 double rows on each side, about 12 in each front double row. 12, 14.
Montagua vermifera.

4 Branchiæ unbranched, situated in very many oblique rows, or in irregular clusters on each side.

Triangular, broad in front; branchiæ oblanceolate, 200-400 in number arranged in 18-24 oblique rows on each side; great variation in size and color from flesh-color to dark olive and brown much mottled.
157, 16. *Æolis papillosa.*

Lance linear; watery white; branchiæ club formed in 7 clusters of about 5 each on each side; back exposed. 12, 2. *Coryphella gymnota.*

5 Branchiæ branching bipinnate, arranged in a circle or crown in the centre of the back.

Ovate-oblong, widest about 1/3 from head;

[illegible] mantle scattered over with fine papillae; mantle does not cover back end of foot; branchial crown of plumes 3/4 from head, retractile into a single cavity; tentacles long, outer half plicated with cross folds, lip smooth, base surrounded with papillae; [illegible] purple, [illegible] white and bright yellow, [illegible] of branchiae [illegible] covered over with bright golden specks. [illegible], 12.

Doris lilida.

Body oblong, ends equally rounded; back covered with large cream colored mushroom like tubercles; wreath of about 7 once pinnate plumes retractile into separate sheaths. [illegible]

Onchidoris pallida.

[illegible] oblong-linear; [illegible] about centre, a [illegible] fringe along the sides of the [illegible] with [illegible] tubercles on each side; [illegible] and [illegible] small bipinnate branchial plumes forming part of circle about [illegible] head nearly circular; yellowish [illegible] with tubercles tipped with sulphur yellow. 15, 6. *Polycera lessonii.*

6 Branchiae in arborescent much branching forms, situated in a single line along each side of back.

Elongated, tapering, rounded, [illegible] high as broad; branchiae [illegible] more pairs translucent; color very variable, pink rose to dark brown mottled and marbled. 66, 157. *Dendronotus arborescens.*

IX Animal without shell; no branchiae; breathing through the entire surface of the body; sides of body dilated into wing like swimming appendages, which are folded on the back while [illegible].

GASTEROPODA.

Swimming appendages when on the back lapping over each other; [illegible] green with white and red spots. [illegible] according to position of appendages.

Elysia chloritica.

Membranous expansion not meeting on the back; head rounded, globose; tentacles short, blunt, broad; [illegible] square in front, pointed behind; Sea green with whitish spots.

6, '25, [illegible] Elysiella catulus.

PTEROPODA.

Family

Free sea mollusks swimming by means of two fin like expansions developed from the anterior extremity. Hermaphrodite. Head mostly rudimentary, expanded into a large muscular fin. Mouth small, sometimes tentaculate. Small active animals, ily colored and mostly provided with thin symmetrical shells and found in large numbers on the surface of the ocean at night.

GYMNOSOMATA. Clionidae

A Animal without mantle or shell; head distinct; fins attached to the sides of the neck.

Foot distinct with a central and posterior lobe; head with tentacular projections.

Gelatinous, pellucid, pale blue; mouth and end of body scarlet out of water; wings somewhat triangular; tail acute; tentacles six, conical; length 32; width including wings 20.

Cliona papillonacea

THECOSOMATA. Hyalaeidae

B Animal with an external shell; head indistinct; foot and tentacles rudimentary, mouth in a cavity formed by the locomotive organs.

1 Shell symmetrical, straight or curved, globular or needle shaped

Shell long, conical, slender, slightly curved towards the acute apex, polished, diaphanous; animal white; wings obovate and bear each a slender tooth near the middle of the ant. edge; length 16, width 8. Styliola vitrea.

Shell long, straight, ant. end dilated, compressed on the sides and terminating

PTEROPODA

[illegible] in a very long spine and with a short spine on each side.
Diacria trispinosa

[illegible] plate[illegible], prolonged[illegible] long pointed [illegible] outward curving [illegible] like processes extending back beyond the tail [illegible] triangular, with the ends [illegible] length [illegible] tail [illegible] of processes [illegible] width [illegible]
Cavolina tridentata

2. Shell minute, spiral, operculate.

[illegible] length [illegible] page [illegible], where it is figured.
Spirialis gouldii

Fig. [illegible]

Cavolina tridentata; natural size.

Fig. [illegible]

Clione papilionacea; natural size.

Styliola vitrea; natural size.

LAMELLIBRANCHIATA.

Fig. 1.

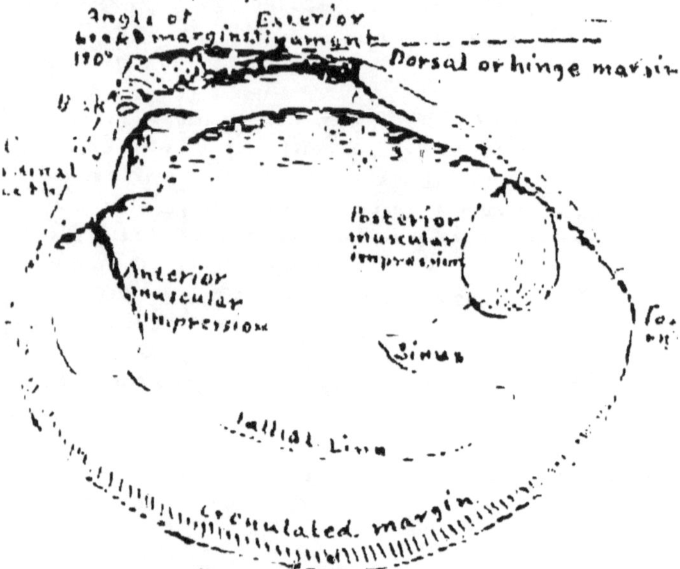

Inside of Right Valve of Venus mercenaria.

Fig. 2.

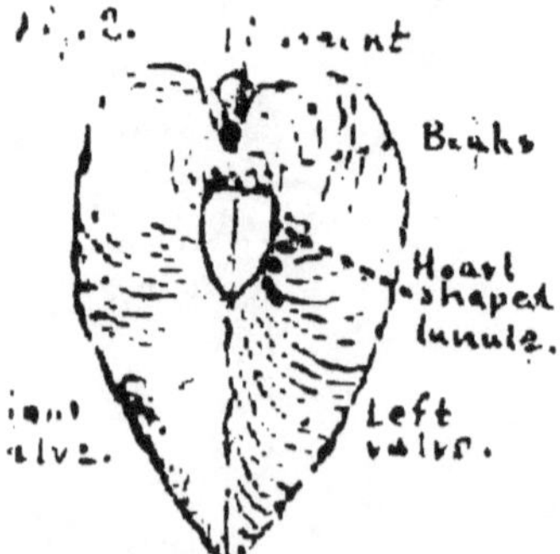

Interior end of
enus mercenaria.

Fig. 3.

Animal and shell of Mulinia lateralis, natural size.

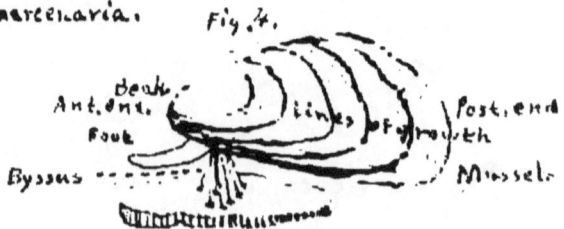

Fig. 4.

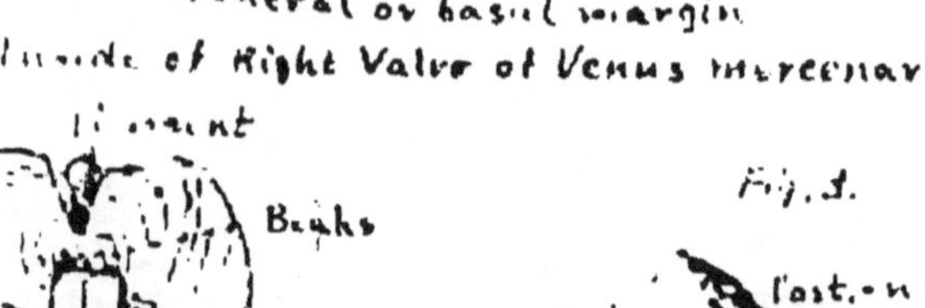

LAMELLIBRANCHIATA

Lamellibranchiata. Headless mollusks encased in a bivalve shell, sometimes with accessory valves. Body enclosed within a mantle. Respiratory organs consisting of lamelliform or filamentous branchiae; Sexes distinct.

The shell though usually equivalve is often inequivalve, but always inequilateral, with usually two but sometimes one adductor muscle for closing the valves.

The shell if hinged is opened by either a ligament outside, or a cartilage inside, or both.

Locomotion is very imperfect in the adult state; many are permanently fixed, either by their shells, or by a peculiar secretion one form of which is known as the byssus, or they bury themselves in the sand, or bore into timber or rocks.

Note.— r.v. = right valve; l.v. — left valve; ant. = anterior, post = posterior, c.t. = cardinal teeth. Numbers representing distance are all in m.m. At the end of the descriptions they are for length in front of beak + length behind beak, height (distance across from hinge to the other edge), breadth (distance from outside of one valve to outside of other), and, if the fifth No. is given, it represents the angle formed by lines joining the beak and the slopes of the shell.

I Shell gaping more or less at the ends.

A Without hinge or ligament often with accessory valves. Animal club shaped or worm like, with short truncated foot; siphons long united to near their tip.

LAMELLIBRANCHIATA.

1 Valves equal, largely open at both ends, forming a ring, placed at the larger extremity of a shelly tube open at both ends, and furnished with pallets.

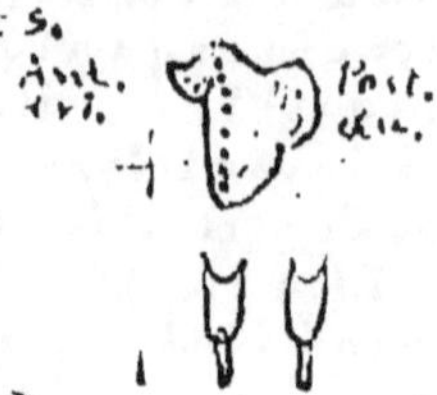

posterior auricle extends down 3/5; the ant. triangle 1/3; pallets emarginate at tip, convex on one side and plain on the other; stalk about as long as the blade (common ship-worm boring in timber).
Valves 5, 5, 5; pallets 5, 1.4; stalk 3/7 of length.
Teredo navalis.

Post. auricle extends a little higher than beak and separated by a narrow notch and extends down 2/3; ant. triangle 1/3 with 20-30 radiating grooves; pallets small ovate rounded or slightly emarginate; stalk pointed. 6, 6, 6; 4, 1.6; — 1/3.
Teredo megotara.

Post. au. and ant. tri. both extend down 1/3; post. au. does not extend up quite as high as beak; pallets battledore like 6, 8, 8; — 1, 1; — 1/2 Teredo thomsonii.

Post. au. rounded not extending above beak but down 3/5; ant. tri. 1/2; pallets small angular obovate blunt.
10, 9, 9; — 6, 2.4; — 1/2 Teredo dilatata.

Post. au. rounded does not extend upward but 1/7 and downward 2/3; ant. tri. 1/3 pallets oar shaped, blade oblong feathered serrated.
6, 6, 6; — 12, 2; — 1/2 Xylotrya fimbriata.

2 Shell gaping equally at ends; a rib like process or tooth arises from the back and shoots nearly across the shell; shell large, more than 40 long; (found in wood, clay and stones which they have perforated.
Shell very large oblong-ovate, white, covered with radiating toothed ribs, the teeth formed by the lines of growth.

LAMELLIBRANCHIATA.

30+100, 50, 50, 180. Phola[illegible]

Shell medium oblong, chalky white; ant. end triangular acute with radiating, irregular ribs; post. broadly truncate; lines of growth distinct.
[illegible]33, 30, 24, 150. Pholas [illegible]

3 Shell oblong-oval, gaping widely at post. end.
Thick and strong; ant. end with triangular with radiating toothed ribs; post. [illegible] and separated from post. end by an oblique broad channel
[illegible] 35, 15, [illegible] [illegible]

one slight exception
B Hinged, equivalve, elongated, 2 or more times as long as high, sub-cylindrical, gaping more or less at both ends, with not [illegible] c.t. in each valve.

* With external ligament

1 Glossy, smooth, thin, with internal rib passing from beak part way [illegible] the shell, often broken off from old shells; beak less than [illegible] to radiating lines.

Shell scabbard shaped; sides nearly parallel, 6 times as long as wide, beak terminal; yellowish or brownish green epidermis
0+150, 25, 20, 80, Ensatella americana
Ovate-elliptical, fragile; beaks minute; light yellow-green epidermis, color blended with livid violaceous in such a manner as to form 2 radiating compartments of each color; rib white inclining backward about ⅓ across shell. 12+32, 17, 7, 160, Siliqua costata.

2 Beak nearly central; sides nearly parallel, more or less curved; ends nearly rounded [illegible]
Oblong with obliquely rounded ends; thick and strong; post. end narrower; beak obtuse and slightly elevated; surface

LAMELLIBRANCHIATA.

coarsely wrinkled by the stages of growth and covered by a dense and strong epidermis which is yellowish in color and folds over the edge, pallial line with sinus which passes beyond beaks.

50+42, 32, 25, 175, Tagelus gibbus.

Oblong-oval arcuated; surface smooth in the central region & wrinkled at the ends, with a band of reddish purple passing from the beaks across the shell growing wider and fainter in its progress, this is visible within and covered by a faint rib like thickening; epidermis straw color.

16+16, 10, 8, 170 Tagelus divisus.

3 Very irregular, about toothless; ligament long and strong; pallial sinus narrow and deep; right valve a little larger than the left; epidermis thin dingy yellow; muscular impressions obscure; foot of animal bright orange yellow; an exceedingly variable rough shell found adhering to almost all kinds of marine objects. 10+15, 14, 10, 160, Saxicava arctica.

4 Oblong-oval, more than twice as long as wide, chalky white with radiating ribs some of them with toothed scales.

Ant. end covered with elevated toothed radiating lines; post end marked with finer radiating lines, 2 small teeth in each valve; distinct ovate lunule.

[illegible]30, 1[illegible], 8, 100, Petricola pholadiformis.

5 Smooth shining thin more or less iridescent with very fine concentric striae; ant. dorsal margin straight or slightly concave; post. end slopes rapidly and is sub-truncate at end; ventral margin nearly parallel with ant. dorsal, color pink, light straw or white often banded concentrically 4+5, 7, 3, 130. Angulus tenellus

LAMELLIBRANCHIATA.

[illegible] With internal cartilage.

Epidermis thick and shining [illegible] extending [illegible] beyond the edge [illegible] [illegible] and [illegible] [illegible] shell [illegible] [illegible] [illegible] [illegible] [illegible] with [illegible] [illegible] lines.

Epidermis [illegible] lines and notches in epid[illegible] [illegible] including epid[illegible] [illegible] 17. [illegible] [illegible] dark brown [illegible] lines and notches 15–20; inside [illegible]ish blue; angles of the [illegible] of epidermis [illegible].

[illegible] 15, 20, 18, 100. [illegible]

[illegible] with [illegible] [illegible] one or both valves [illegible] [illegible] the shell, [illegible] no cardin[illegible]

[illegible] valve [illegible] [illegible] long.

Small, ovate globose, rounded before [illegible] pointed behind; a distinct [illegible] running [illegible] post. end to [illegible] [illegible] [illegible] [illegible] white [illegible]

[illegible] and [illegible] [illegible] [illegible] [illegible] [illegible] [illegible] [illegible] truncated [illegible] [illegible] [illegible] [illegible] the shell and [illegible] [illegible] [illegible] [illegible] [illegible]

2 [illegible]-equivalve no V [illegible]

Narrower and a little more pointed; dingy white covered by a dirty brown epidermis; spoon shaped process extending from the left valve into the right; very common. (soft clam). 43+47, 50, 25/60. Mya arenaria.
Thin, fragile, pearly, translucent, oblong-ovate, slightly gaping at one end, post. truncate, elongated; beaks prominent, inclined forward; process consisting of a narrow ledge within each valve; epidermis projecting beyond the edge of the shell and wrought into irregular fringed wrinkles often containing grains of sand. 1+11, 11, 5, 115. Lyonsia hyalina.
Small, thin, fragile, bluish white, ovate-triangular; spoon shaped process shallow, a single oblique tooth by its side in each valve; pallial line with large sinus; lines of growth shown by raised ridges; shell warped; ant. end rounded, post. more pointed. 1+1, 11, 5, 110. Cumingia tellinoides.

3 Equivalve, with extra V tooth; thick and strong, large ovate, with dirty brown epidermis, spoon shaped cavity very large and broad, the V tooth very delicate. 40+65, 70; 45, 110. Mactra solidissima.
Small triangular; spaces before and behind beaks broad heart shaped; nearly smooth shining, with a thin dirty brown epidermis; the pits are small and deep and before it is a strong V tooth. 5+8, 11, 5, 11. Mulinia lateralis.
Sub triangular; smooth and covered by a shining golden yellow epidermis; spoon shaped pit very deep; a long acute V teeth. 1+10, 25; 14, 120. Cronia ...

D Hinged, equivalve, minute; tooth on the shorter side excavated for the cartilage in place of a perfect spoon shaped cavity.
Ovate, fragile, white within and without; 1 teeth in each valve; nearly closed, beaks elevated nearly central; surface shining with a very thin straw colored epidermis; margin polished but marked faintly with

Anatinidæ. Tellinidæ. Mactridæ.
SIPHONATA.

LAMELLIBRANCHIATA.

[illegible], [illegible] 290, [illegible], 190
Monterosato [illegible]

1 [illegible], inequivalve, both valves [illegible] than the right; length [illegible] the height; [illegible] in each [illegible].
1 Right valve flat.

[illegible]

[illegible] rounded, [illegible] and narrowing [illegible] hinge margin [illegible] lines, 101-6, [illegible]

Clidiophora trilineata

2 Not very inequivalve, small, less than [illegible], oval to oblong, thin; outside ligament iridescent, very thin; ligament short and prominent; varies from white to a slight tinge of rose color; [illegible] 2 to each valve. 513, 8, [illegible], 150. Angulus tener

Iridescent, smooth, shining, white pink or light straw color; ligament plate longer; ant. dorsal margin nearly straight or slightly [illegible]. 9[illegible], 3, 146. Angulus tenellus

[illegible] and widely gaping; lines of growth fine; inside polished and with fine radiating lines; hinge delicate, two diverging teeth in [illegible]. 9[illegible], [illegible], 145. Tellina tenta

1 [illegible], inequivalve, [illegible] more convex than the left; length no more than 1½ the height; beaks conspicuous; the right one excavated to receive the tip of the left; outside ligament large and protuberant.

Shell large, thin, light and fragile, [illegible] dingy white color; [illegible] teeth; [illegible] surface coarsely wrinkled [illegible] the lines of growth and [illegible] by a ridge running from beak to [illegible] angle. [illegible], 05, 36, [illegible]. Tamoeis [illegible]

LAMELLIBRANCHIATA.

Small, white, not very thin; hinge callosity not spoon shaped, ovate-triangular, beaks post. ant. dorsal margin nearly parallel with ventral. 12+6, 12, 7, 180. Interior clear white. *Thracia truncata.*

G Shell equivalve and very slightly gaping at ends; not more than 2 c.t. to each valve; 1½ to 1¾ times the height equals the length; Animal with long slender siphons; mantle open.

Pallial impression with deep sinus.

Ligament external.

Ovate-orbicular; muscular impressions distinct, varies much in size, solidity from thin to moderate and color from white to bluish or rusty with dark epidermis; very common. 10+12, 17, 9, 135. *Macoma fragilis* var. *fusca.*

Sub-oval, thin, brittle, white, covered with dusky epidermis; ant. end rounded, post. angular; sinus nearly reaching ant. muscular impression; interior bluish white. 6+9, 16, 6, 140. *Macoma tenuis[illegible].*

H Shell equivalve; slightly gaping at the post. end; neither ribbed or spinous; length and height equal; c.t. not over 8.

Small, thin, sub-[illegible], smooth, pale fawn color, sometimes [illegible] with dark brown; within [illegible] yellow, with a purplish blotch at the post end; epidermis thin and of a darker color. 8+13, 20, 16, 110. *Laevicardium mortoni.*

I Cardinal teeth small, numerous, disposed in a line along the hinge margin.

Oblong-ovate or kidney shaped, somewhat pointed in front, broadest and truncate behind gaping at both ends; a rib like

45°. 23 + 34, 35, 25,
See* page 43

A

1.
curving

2 No lunule;

car. t. epider

n

[illegible]IBRANCHIATA.

3 Shell heavy; no lunule; 3 diverging c. t. in each valve; no sinus.

Oval pit in place of lunule; epidermis of a dark shining brown color; inside chalky white, no purple
23 60, 75, 45, 120. *Cyprina islandica.*

B Shell small to medium, [illegible]  shape; with [illegible] radiating ribs or ridges; [illegible] or [illegible].

1 Hinge with a single [illegible] teeth under the beaks.

Thick and strong; height greater than length; surface with [illegible] radiating [illegible] which are crossed [illegible] distinct lines of growth, and covered by a strong rusty-brown epidermis; lunule deep, [illegible]; ligament small and [illegible] [illegible] 17, 18, 141, *Cyclocardia [illegible]*.

[illegible] than length; about [illegible] ribs; smoothish. 8 + 14, 18, 12, 120.
Cyclocardia [illegible]

2 Hinge with numerous small teeth in [illegible] separated by a diamond [illegible] area; [illegible] radiating ribs [illegible] in [illegible].

Oblong; beaks prominent, directed very obliquely forward and terminating nearly over ant. end of [illegible] of c. t.; [illegible] [illegible] epidermis. 12 + 52, [illegible], 120. See +++ Page 44

[illegible] not very oblique and terminating about over ant. third of row of [illegible]; [illegible] but slightly curved.
[illegible] See ++ Page 44.

LAMELLIBRANCHIATA.

2. Small; [illegible] oblique [illegible] to [illegible] [illegible] sinus; no lunule.

Fragile, dingy white; about [illegible] radiating ribs; these are slightly rounded and have a series of equidistant [illegible] scales; interior [illegible] of flesh color, [illegible] furrowed to suit the ribs outside. [illegible] 10, 7, [illegible] *Cardium pinnulatum.*

[illegible] Small, thin, rounded oval, regularly convex; lines of growth minute, [illegible], minute and crowded radiating lines, which increase in [illegible] toward the margin; margin crenulated.
Epidermis [illegible] yellow; teeth [illegible] not in contact. This shell belongs [illegible] [illegible] mussel, and the [illegible] [illegible] to the shell in the [illegible] [illegible] 10, 10, [illegible], 120. [illegible]

[illegible] minute to medium, smooth, length and height nearly equal; distinct lunule.

Minute, dead white, [illegible] thin; teeth prominent inclined [illegible] lunule [illegible] slightly indented; [illegible] with a remarkable widened [illegible] which runs near the margin to the post. [illegible] producing a loop indentation in the outline; [illegible] [illegible] [illegible] [illegible] [illegible]

[illegible] [illegible]

Minute, [illegible] [illegible], white, rather thick, [illegible] [illegible] [illegible] [illegible] lines of growth; [illegible] [illegible] central teeth, [illegible] [illegible]

[illegible] slightly and slightly [illegible]
[illegible] light yellow [illegible]
[illegible] bluish [illegible] l.v.
[illegible] *Astarte quadrans*

[illegible], thick and [illegible]
[illegible] [illegible]
[illegible] much elevated, [illegible]
[illegible]-lanceolate, [illegible]
[illegible] smooth and [illegible]
[illegible] chestnut, wrinkled [illegible]
[illegible] with black [illegible]
[illegible] small; [illegible]
[illegible] *Astarte castanea*

[illegible] to medium, with [illegible]
[illegible]
[illegible] very distinct. [illegible]

with distinct [illegible] and [illegible]
in [illegible] each valve.

[illegible]
[illegible], with [illegible]
[illegible] rounded [illegible]
tric rib [illegible] depressed [illegible]
wider than the [illegible]
long, narrow, [illegible]
[illegible] polished [illegible]
f.c.t. in [illegible] l.v. 12+18, 25 [illegible]
Astarte undata

LAMELLIBRANCHIATA.

Small, pointed, inclining much forewards over a small lanceolet lunule; surface covered with remote concentric lamellar ridges, with several thread like striae between them; minute radiating lines near margin; interior chalky white except near margin where it is polished; 2 small c.t. in r.v., l in l.v., large.
1C+24, 40, 15, 140. Lucina filosa

Orbicular, lenticular, thin, white, glossy; beaks elevated and slightly inclined forward; stages of growth strongly marked and sculptured with regularly disposed remote and nearly parallel lines which bend at nearly a right angle from the centre of the shell and pass obliquely downward towards the ends of the shell forming teeth around the entire margin; lunule [illegible] narrow; l c.t. in r.v., 2 small in l.v.
1:+13, 24, 15, 1[illegible]0. Cyclas dentata.

Small, quadrant shaped; ant. margin straight or concave; about 14 concentric ridge like waves, which have minute [illegible] radiating lines between them; lunule long and deep; color pale yellow-green with dusky markings.
[illegible] Gouldia mactracea.

Small, white, [illegible], higher than long concentrically striate, much swollen in the middle; beaks are prolonged and turned strongly to the ant. side; lunule large and sunken, somewhat flat; int. border with a prominent rounded angle; post. side with two strongly developed flexures separated by [illegible] ves; inside with radiating grooves near the ventral edge.
7+8, 13, 13, 100. see [illegible] 19.

LAMELLIBRANCHIATA.

2 Without lunule; c.t. less than 5 in each valve; minute nearly orbicular.

Three c.t. in l.v., middle one conic triangular, two in r.v.; sinus narrow; ant. end more rounded; ant end and most of the base white tinged with rose color, post. and upper portion reddish purple; within white except post. which has the purple of the outside; common. 1¾+2, 3, 1.6, 110. Tottenia gemma.

Post. more rounded; shell somewhat triangular; rather solid; shining straw color; no purple; sinus very small. 1½+1½, 3, 1.6, 95. Tottenia manhattensis.

3. Small to minute; a series of eight or more c.t. in each valve; no sinus; a spoon shaped pit for the cartilage.

Minute, somewhat triangular; oblique globose; 3 ant. t. and 7 post. t. in each valve; epidermis olivaceous; coarse unequal lines of growth.
3+1, 3½, 3, 90. See ++ page 43.

Trapezoidal, thin, 8 ant. t. and 4 5 post. t., post. ones long and slender, interior silvery white but not pearly; epidermis grass green; beaks prominent; no radiating lines
6+1½, 6, 4, 110 See + page 44.

Oblique ovate-triangular; ant. end short, 12 ant. and 18 post. teeth, some very small near spoon shaped pit, the series of teeth nearly at right angles to each other; interior pearly; epidermis light olive color, with darker zones.
2+8, 8, 6, 85. See + page 43.

LAMELLIBRANCHIATA.

E' Elongated shell, at least [illegible] as long as high, without sinus [illegible] without siphon; two adductor [illegible] [illegible] one [illegible] impressions [illegible]

[illegible] smooth, green or greenish yellow [illegible] epidermis; [illegible] many [illegible] valve, pit for cartilage [illegible] [illegible] of teeth.

[illegible] by [illegible] tip, [illegible] and light green epidermis [illegible]

Interior bluish white; [illegible] the length of the shell, [illegible] post. in each valve; cartilage pit small.
[illegible] 5, 10, 11, 170. Yoldia li[illegible]

Interior pearly white; [illegible] cartilage deep and triangular [illegible] each valve.
[illegible] Yoldia [illegible]

[illegible] elongated oval; [illegible] [illegible] green epidermis; [illegible] each valve. [illegible] 140
[illegible]

[illegible] 7. Yoldia thracia form [illegible] [illegible]; pit for ligament [illegible] [illegible] epidermis [illegible] in each valve.
[illegible] 6, 170. [illegible]

[illegible] Leda [illegible]

[illegible] Nucula [illegible]

LAMELLIBRANCHIATA

† From 61 Nucu[illegible]

†† From 88 Scapha[illegible]

††† From 98 Argi[illegible]

†††† Arca [illegible]

2 Triangular [illegible] or nearly [illegible] ling along most of the [illegible] adheres by a byssus [illegible]

[illegible] triangular [illegible] straight, [illegible] violet [illegible] inside [illegible] violet. [illegible]

[illegible] margin ascending [illegible] the length of the shell, [illegible] arched a little up[illegible] dark chestnut colo[illegible] roughly marked by lines and minute radiating [illegible] angle of [illegible]

[illegible] surface [illegible] radiating well marked [illegible] striae or indented lines [illegible] green and yellow [illegible] shell [illegible] angle of [illegible]

LAMELLIBRANCHIATA.

Rhomboidal to ovate, with numerous radiating striae, in two sections, leaving the middle portion smooth.

Small, thin, long ovate; surface beautifully sculptured with a network of very minute lines of growth and very numerous fine indented radiating lines which are obsolete on 1/3 of base at centre; epidermis rusty brown; rare. 3+12, 9, 6, angle of margins 45. Modiolaria nigra.

Ovate-oval, hinder extremity somewhat lobed; ant. end with about [illegible] post. with many radiating lines; the limit between the post. lines and the middle smooth portion marked by an elevated ridge passing from beaks; epidermis olive green with chestnut shades; interior brilliantly silvery. 3+21, 15, 10, angle of margins [illegible]. Modiolaria [illegible].

Irregularly oval (heart shaped from front); ant. end with 11 or more ridges, post. with many; epidermis greenish yellow with shades of olive; within silvery; [illegible] crenulated. 15, 11, 7, 6, angle of margins 16. Modiolaria corrugata.

[illegible] Crenella glandula.

I Shells with single adductor muscle

II Elongated shell, irregularly [illegible] [illegible] larger valve [illegible]

LAMELLIBRANCHIATA.

hering;
large narrow gradually widening from hinge end; (the oyster)
Ostrea virginiana.

2 Orbicular; more or less inequivalve; hinge line straight

Lower valve nearly flat and nearly smooth; upper valve without ribs; diameters about 150, breadth 35. Pecten tenuicostatus.

With 50–100 radiating ribs; length 75, height 85, breadth 25.
Pecten Islandicus.

About 20 radiating ribs, l. 65, h. 62, b. 50. Pecten irradians

3 Orbicular, irregularly foliated, very inequivalve, under valve nearly flat, perforated near the beak for passage of muscle by which it adheres

Surface rugged, scaly, variously wrinkled; aperture ovate; beak not quite reaching margin; pearly; usually 25 in diameter
Anomia glabra.

Beak reaching margin; upper valve with fine prickly radiating lines; lower valve smooth; aperture circular; color yellowish white; diameter 12. Anomia aculeata.

www.ingramcontent.com/pod-product-compliance
Lightning Source LLC
Chambersburg PA
CBHW022026190326
41519CB00010B/1617
* 9 7 8 3 3 3 7 3 3 0 7 9 8 *